# Going Metric the Fun Way

# Going Metric the Fun Way

- Riddles
- Tricks
- Teasers
- moron jokes
- Cut-ups
- Upside Downers
- funny facts
- Homemade Metric Measures

by Eva-Lee Baird and Rose Wyler
Illustrated by Tālivaldis Stubis

Doubleday & Company, Inc., Garden City, New York

**Library of Congress Cataloging in Publication Data**

Baird, Eva Lee.
 Going metric the fun way.

 Includes index.
 SUMMARY: Introduces the principles of the metric system through games, activities, puzzles, and riddles.
 1. Metric system—United States—Juvenile literature. [1. Metric system] I. Wyler, Rose, joint author. II. Stubis, Talivaldis, 1926– III. Title. QC92.U54B34 530′.8

ISBN: 0–385–13641–2 Trade
ISBN: 0–385–13642–0 Prebound

Library of Congress Catalog Card Number 77–16895
Text copyright © 1980 by Eva-Lee Baird and Rose Wyler
Illustrations copyright © 1980 by Talivaldis Stubis
All rights reserved
Printed in the United States of America
First Edition

## CONTENTS

**FOREWORD** ............................................................. 6

**GETTING IT ALL TOGETHER** .......................... 7

Homemade metric measures / Seeing yourself metrically / Puzzlers / Show-off stunts

**MORTON THE MORON STARTS HIS CAREER** ............................................................. 31

You know Morton. He's ambitious. That's why he tries to learn the metric system.

**SQUARE SHOOTERS** .......................................... 45

Teasers with straight, true-blue answers, although the questions are a little tricky.

**KLONDIKE MIKE AND THE SOLID THAT COULD BE POURED** ........................................ 69

A teaser tale about a gold miner in the bad old days.

**WHEN YOU'RE LOOKING FOR SOMETHING TO DO** ................................................................ 81

Recipes / Arty crafts / Upside downers / Zany puzzlers / Clever gadgets calling for metric know-how.

**JEST RIDDLES** ...................................................... 101

Puns / Cornball / Zanies. And if that's what you like, these riddles are jest for you.

**INDEX** .................................................................... 128

**FOREWORD**

We did it! The 200-year argument is over. The United States is switching to the metric system, the most modern, streamlined measuring system yet devised.

The system is so easy to use. There's none of this business of 12 inches to a foot, 3 feet to a yard, and 5,280 feet to a mile. Instead, you just remember that 100 centimeters equals 1 meter and 1000 meters equals 1 kilometer. No tricky arithmetic is needed to handle units of differing sizes.

Yet there's a trick to learning the metric system. Try to avoid switching back and forth between the old and new measurements. That causes confusion. Start using the metric system only, and stick with it. It's as simple as that.

# Getting it all together

You don't have to buy any metric measures or rulers, for you can make your own. Here are the directions.
You can go metric now—and step into the future.

For your first step, try using a tool from the past to measure centimeters. Strange though it seems, this tool is just like one that the ancient Egyptians used. Any idea what it is?

. . . the human hand. It's a very ancient ruler—and it still works.

Spread out your fingers as wide as you can and you have before you one of the oldest measures in the world. This is the hand span—the distance between the tip of the little finger and the tip of the thumb.

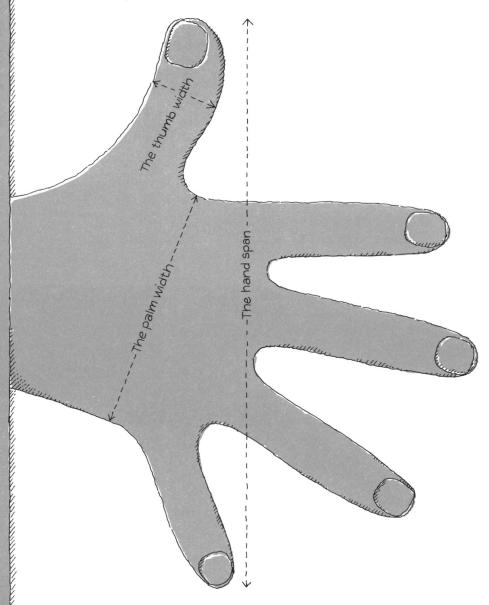

Ancient records show that the Egyptians used the span, but they give no idea of the size of this measure. Maybe it had no definite size. In measuring a piece of cloth, Sheba might say it was 8 spans long, while Ahmed, whose hand was bigger, would say it was only 7 spans.

It's easy enough to convert the hand span into a metric measure. Just use the ruler on the edge of the page. It is marked in centimeters, a unit of length in the metric system. Place the tip of your thumb on the zero mark of the ruler and see how far you can make your little finger go. This gives you your hand span in centimeters.

The width of the thumb is a convenient measure too. So is the width of the palm. Make these measurements, then remember them, and you have a secret ruler that is never more than an arm's length away.

## MYSTERY MEASURING

Challenge people to guess the size in centimeters of some familiar object. A book or a dish or a leaf will do. Most people won't be able to guess accurately, but you know the answer, because . . . .

... you measure the object with your secret ruler.

### THE MAN WHO FORGOT HIS FEET

An elderly man in Maine had been in an accident, and as a result, he suffered a loss of memory. He completely forgot inches, feet, and the other measurements he had used all his life. Instead of worrying about this, he decided to go metric.

He quickly learned to use centimeters and meters, and he enjoyed making measurements with his new metric tape. But he kept forgetting where he put the tape. He needed something to use in its place—something he could never lose.

It didn't take long to find a substitute for the tape. Several were found, in fact. And guess what they were. . . .

... His feet, legs, arms, and hands became the substitutes.

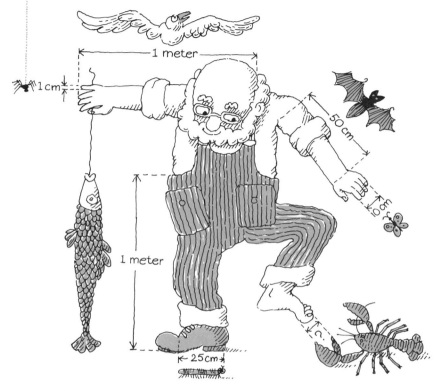

The old man had no trouble remembering his measurements. He was so pleased with himself, he pasted a bumper sticker on his car that read:

## FREE TAPES

It's really a great convenience to use different parts of your body as metric units. But how can you measure your arms and legs without buying a centimeter tape?

... Let someone else buy the tape. That's a fine answer for a riddle. But in real life you might have to get a tape some other way. Consider using a homemade one.

Here are the directions for making a tape measuring 200 centimeters. That's long enough for most purposes.

> Take a sheet of typing paper 28 centimeters long. Using the ruler on page eight, make guide marks 3 centimeters from the bottom. Then draw a line across the sheet.
>
> Fold the sheet lengthwise, first in half, then in quarters, then in eighths. This will give you eight strips, all the same width. Cut along the folds.
>
> Place a strip on something flat and firm. Put a second strip on bottom so that its top overlaps the last three centimeters, and paste it in place.
>
> Repeat this with the rest of the strips. Cut off the bottom 3 centimeters from the last strip and you have one long tape, measuring 2 meters.
>
> Mark the tape in centimeters, using the ruler on page eight as your guide. Put the bottom end on the 0 line, and mark off 20 centimeters. Then move the tape so that the 20 mark is over the 0 mark of the ruler and check off the intervals from 20 to 40. Keep on going until you reach the end of the tape. Now it is ready to use.

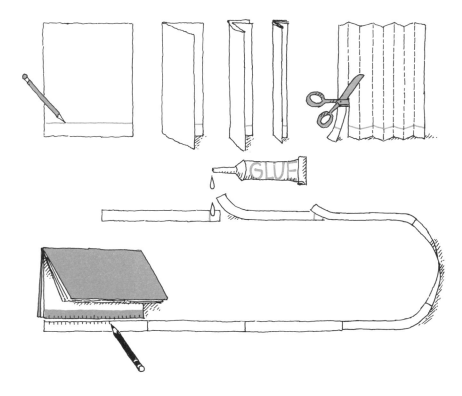

## SHORT AND TALL—MEASURE THEM ALL

A homemade centimeter tape comes in handy in measuring people's height. Paste the tape on a doorjamb. Make sure the 0 mark is right on the floor, and you're all set.

Whose height will you measure first? Your own? Take off your shoes and stand with your back against the doorjamb. Hold a piece of cardboard on top of your head, placing it so the edge touches the tape. Then turn around and see how tall you are.

How tall are your friends? You probably won't find anyone taller than 200 centimeters—unless you know some basketball players.

## DO YOU SEE YOURSELF HERE?

Now that you know your height in centimeters, you will want to know your metric weight. The unit for this is the kilogram, which is equal to 2.2 pounds.

Weigh yourself on a kilogram scale, if you can find one. If you can't, get your weight in pounds, then convert it to kilograms. A calculator will help, since a pound equals .454 kilogram. To convert, multiply the number of pounds you weigh by .454, or just use this table:

| pounds to | kilograms |
|---:|---:|
| 30 | 13.6 |
| 35 | 15.9 |
| 40 | 18.1 |
| 45 | 20.4 |
| 50 | 22.7 |
| 55 | 24.9 |
| 60 | 27.2 |
| 65 | 29.5 |
| 70 | 31.8 |
| 75 | 34.0 |
| 80 | 36.3 |
| 85 | 38.6 |
| 90 | 40.8 |
| 95 | 43.1 |
| 100 | 45.4 |
| 105 | 47.6 |
| 110 | 49.9 |
| 115 | 52.2 |
| 120 | 54.4 |
| 125 | 56.7 |
| 130 | 59.0 |
| 135 | 61.2 |
| 140 | 63.5 |
| 145 | 65.8 |

Are you in good shape—not too fat, not too thin? These charts show the ideal range of weight for your height and age. Use it to check your measurements, and you will find that going metric can be a very personal affair.

**BOYS AND GIRLS**

| height in cm (without shoes) | weight in kg |
|---|---|
| 90 | 12.0 — 14.2 |
| 95 | 13.1 — 15.3 |
| 100 | 14.2 — 16.8 |
| 105 | 15.5 — 18.3 |
| 110 | 17.0 — 20.0 |
| 115 | 18.6 — 21.8 |
| 120 | 20.3 — 23.7 |
| 125 | 22.5 — 26.1 |
| 130 | 24.9 — 28.7 |
| 135 | 27.7 — 31.7 |
| 140 | 31.0 — 35.2 |
| 145 | 34.6 — 39.0 |
| 150 | 38.8 — 43.4 |
| 155 | 43.1 — 47.9 |
| 160 | 47.5 — 52.5 |
| 165 | 51.9 — 57.1 |

**TEEN-AGERS AND ADULTS**

| height in cm (without shoes) | weight in kg | |
|---|---|---|
| 155 | 53.5 — 58.5 | Male |
| 160 | 56.2 — 61.7 | |
| 165 | 59.0 — 64.9 | |
| 170 | 62.6 — 68.7 | |
| 175 | 66.2 — 72.6 | |
| 180 | 69.9 — 77.1 | |
| 185 | 73.5 — 81.6 | |
| 190 | 78.0 — 86.2 | |
| 145 | 44.5 — 49.9 | Female |
| 150 | 47.2 — 52.6 | |
| 155 | 49.9 — 55.3 | |
| 160 | 52.8 — 59.0 | |
| 165 | 56.2 — 62.8 | |
| 170 | 59.9 — 66.7 | |
| 175 | 63.5 — 70.3 | |
| 180 | 67.1 — 73.9 | |

### WEIGHT GUESSING—THEN AND NOW

Crowds used to gather around the weight guesser at the old-fashioned country fair. His scale was always busy. Customers had to pay only if he guessed their weight correctly—and most of them paid. Usually he came within two pounds of a person's weight.

That kind of skill takes years to develop, and no one bothers learning it now. Weight guessing, these days, is just a stunt used at parties and club meetings. The modern expert makes guesses in kilograms instead of pounds, and gets correct results by using trickery instead of judgment. The audience realizes this sooner or later, and then they are asked to figure out how the trick works.

Maybe you can figure it out too. Here's the setup. The scale is an ordinary bathroom scale. If it doesn't show weight in kilograms, a conversion table is pasted on it. An assistant sits near the scale, ready to check the readings. During a weighing, the expert stands some distance from the scale, then tries to guess the result. And each guess is right because . . .

... the assistant, sitting near the scale, signals to the expert in code. He arranges his feet a certain way for the first digit, then he sits back and smiles and rearranges them for the second digit, using this code:

## FROM QUARTS TO LITERS

Was the milk you drank today poured from a liter carton? This measure is rapidly taking the place of the quart, which doesn't fit in with the metric system. The liter does. It is based on a metric unit for volume: the cubic centimeter.

A cubic centimeter takes up the same amount of space as a cube one centimeter in length, width, and thickness. A liter takes up a thousand times as much space. In other words, it equals 1000 cubic centimeters.

How does it compare with the quart in size? The liter is a little bigger. It equals 1.05 quarts, which means it holds an extra 10 teaspoonfuls. The two measures are so close in size that liter and quart milk cartons look very much alike. Yet you can quickly tell which is which. All you have to do is . . .

. . . read the label, or printed matter, on the carton.

The law requires that the size be clearly marked on any sealed container used in selling milk.

### ELEPHANTS AND BLUEBERRIES

You probably know the old joke: What's the difference between an elephant and a blueberry?

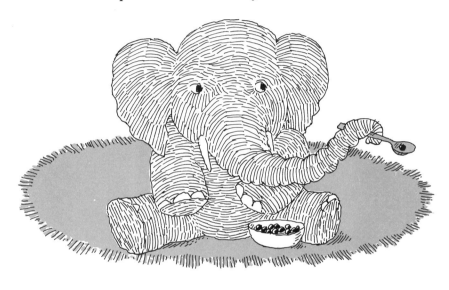

. . . The answer is that the elephant is gray.

Of course, elephants and blueberries are different in size, too. It would be a lot of trouble using the same unit of length to measure them. And it would be unnecessary. Both the customary and the metric system have units of many sizes. There are units to use in measuring the thickness of a sheet of paper, and units for the distance to Mars.

If you use the customary system, you may have to do a lot of figuring to change a small unit into a larger one. In measuring length:

```
  12   inches   = 1 foot
   3   feet     = 1 yard
  5½   yards    = 1 rod
  40   rods     = 1 furlong
   8   furlongs = 1 mile
```

To get the number of inches in a mile means multiplying 12 × 3 × 5½ × 40 × 8. Even with a calculator, it takes time to find that the answer is 63,360 inches.

That kind of problem is easy in the metric system. The steps between the units go up by multiples of ten, as this table shows:

```
10 millimeters  = 1 centimeter
10 centimeters  = 1 decimeter
10 decimeters   = 1 meter
10 meters       = 1 decameter
10 decameters   = 1 hectometer
10 hectometers  = 1 kilometer
```

To find the number of millimeters in a kilometer, you just multiply 10 × 10 × 10 × 10 × 10 × 10, which comes to 1,000,000—a million.

Now back to the blueberry and the elephant. Which units would you choose in measuring them? People hardly ever use the decimeter, the decameter, or the hectameter. So they're out. And the kilometer is too big. You probably would choose the millimeter or the centimeter for the blueberry, and the meter for the elephant. A large blueberry is about a centimeter wide; a large elephant, from trunk to tail, is about 5 meters.

The kilometer is generally used for distances between places. Larger units can also be used. The metric system provides three of them, each a thousand times larger than the one before it. There's the megameter, equal to 1,000 kilometers, the gigameter, equal to 1,000 megameters, and the terameter, which is 1,000 gigameters. But these units are rarely used. Even astronomers, who measure the enormous distances in outer space, avoid them. The reason for this is . . .

... it's easier to use the same unit in comparing distances from place to place on the earth and within the solar system. The kilometer is a convenient size for this purpose, since the distance from the sun to the farthest planet, Pluto, is about six trillion kilometers.

For the vast distances beyond the solar system, scientists use a special astronomical unit called the light-year. This is the distance light travels in a year—nine and a half trillion kilometers. Any unit smaller than that would be of little value, for many stars are over a million light-years away.

### MINI MEASURES

The smallest unit of length used in everyday life is the millimeter. This unit is marked off on centimeter rulers and meter sticks. Although it is small, it is not small enough for some kinds of scientific work. When measuring atomic particles and wavelengths of light, physicists use a special unit called the angstrom, which is only a ten-billionth of a millimeter. Another unit, which is not quite so small, the micron, is used to measure viruses, bacteria, and cells. As its name suggests, it is a convenient size for measuring objects seen under a microscope. It is a thousandth of a millimeter.

Instruments with micron markings are very expensive precision tools. Yet some things of micron size can be measured with an ordinary centimeter tape. If you do some arithmetic along with the measuring, you can find the thickness of a sheet of paper with the ruler on page 8, or with a homemade centimeter tape. Can you figure out how to do this?

... Instead of measuring one sheet, you use a pile of sheets. Stack them up until the pile is one centimeter thick, then count them. If you have 125 sheets in the pile, the thickness of a single sheet is $1/125$ of a centimeter, which is .008 cm when written as a decimal. Since 10,000 microns equals one centimeter, you multiply the fraction by 10,000. Here's the arithmetic:

$$1/125 \text{ cm} = .008 \text{ cm}$$
$$.008 \text{ cm} \times 10,000 = 80 \text{ microns}$$

## ERRORS COME IN DIFFERENT SIZES

Suppose an astronomer measures the distance from the earth to the moon and finds the measurement is wrong by one meter. Another scientist measures a louse egg and finds that the measurement is off by one millionth of a meter—one micron. Which measurement is more accurate?

*. . . The astronomer's work is more accurate.*

In rating a measurement for accuracy, the error is compared to the whole measurement. It is treated as a fraction of the total. The smaller the fraction, the smaller the error.

The length of a louse egg is about 1 millimeter (1000 microns). So an error of one millionth of a meter (1 micron) comes to one thousandth of the measurement. This is a fairly small error.

Now let's rate the moon measurement. The moon's distance from the earth is about 385,000 kilometers; that is, 385,000,000 meters. An error of one meter amounts to less than a three hundred-millionth part of the total distance. It is an extremely small error—one that is 100,000 times smaller than the error in measuring the louse egg.

## METRIC RUNDOWN

Add or take away zeroes . . . that's all you do in changing the size of a metric unit of length, weight, or volume. The arithmetic is so simple because only multiples of ten are used in forming larger and smaller units.

The name of each unit explains its size. When the same multiple is used with meter, gram, and liter, the same prefix is used too. The prefix kilo stands for 1000. So the word kilometer means a thousand meters, just as a kilogram means a thousand grams.

In all, there are sixteen prefixes for units that range in size from a quintillion times the basic measure down to a quintillionth of it. An international committee of metric experts worked out the prefixes, along with a set of symbols for them. But the experts outdid themselves. Many of the prefixes have never been used. Only three appear in the names of common measures, so these are the only ones you need to remember.

**THE COMMON METRIC UNITS**

| Unit name | Symbol | |
|---|---|---|
| meter | m | |
| gram | g | |
| liter | L | |

| Prefix | | Meaning multiply by . . . |
|---|---|---|
| kilo | k | 1000 |
| centi | c | 0.01 (one hundredth) |
| milli | m | 0.001 (one thousandth) |

Notice that all the symbols listed are small letters except one. The liter symbol is a capital. Why this exception?

. . . A small l looks like the number 1. If used, 1l—that is, one liter—can be confused with the number 11. But until recently the small l was the symbol.

Styles of writing and talking about metric units are not set once and for all. They have changed from time to time, and as new needs come up, they will change again.

## HOME ON THE RANGE

One of these days, Texas ranchers will take to wearing 40-liter hats. Sure as shooting. But will they ever tell you how many hectares of land they own? The hectare is the metric unit used in measuring land. It equals 10,000 square meters, which is about 2½ acres. The abbreviation for it is ha.

Maybe haha will come to mean two hectares. And maybe not. Perhaps people will want to stick to acres. Holdings seem bigger that way. A Texan who owns 500 acres may not like saying his ranch is about 200 hectares. If he likes big numbers, and is metric-minded, he will prefer to say that he owns . . .

... 2,000,000 square meters.

And if he wants to use metric shorthand, he'll write m² instead of square meters. This is the up-to-date way to write units.

Cm² means square centimeters; mm², square millimeters. So if you see 4 cm², read it as four square centimeters. If you see 4 mm², read it as four square millimeters. When a unit is used to measure three dimensions, a small three is used. One cubic centimeter is written 1 cm³; one cubic millimeter, 1 mm³.

But trust the Texas ranchers. Even if he measures a coyote's teeth in millimeters, he'll do it in the same old way . . . very carefully.

### TRUTH IS STRANGER THAN SCIENCE FICTION

Anything that takes up space can be measured. All scientists agree on this point. It also goes without saying that they use the metric system.

Yet it is a fact that some of the substances take up space have neither size nor shape. Have you any idea what they are? And can you figure out what kind of unit is used in measuring them?

There are also substances that have size but no shape. Can you figure out what they are and what kind of unit is used in measuring them?

... The substances that have no size or shape are gases. They spread out and take up all the space they can. Each kind of gas can be weighed. A small quantity, a very fine balance, and milligram weights are generally used to do this.

... The substances that have size but no shape are liquids.

Milk in a jar has the shape of the jar. It has no shape of its own. Pour it into a carton and it fits into the carton corners. Its shape changes, but its size does not. Its volume remains the same.

... Volume is usually measured in liters. If the quantity is small, however, volume is measured in cubic centimeters ($cm^3$) or cubic millimeters ($mm^3$).

### REAL, BUT WITHOUT SUBSTANCE

The metric system also has units for things that exist although they have no substance. Perhaps you can figure out how to measure:

1. Something that doesn't take up space, yet can be measured.
2. Something that doesn't take up space, yet increases and decreases.
3. Something that can't be lifted, but can be used in lifting thousands of things.

1. Time. In the metric system, it is measured in seconds. Minutes and hours are not used in scientific work, because they do not increase by multiples of ten, as metric units do. The symbol used for the second is s.

2. Temperature. This is measured in Celsius degrees. On the Celsius scale, zero is the freezing point of water; 100 degrees is the boiling point. These points are written 0°C and 100°C. Note the capital C. That's because the unit was named after a person: Anders Celsius, the man who invented the scale bearing his name.

Weather Bureau reports are often in Celsius degrees now. If you must convert to Fahrenheit to understand them, multiply the temperature in Celsius by 1.8 and add 32. But try to get along without converting.

3. Electricity. Of course you know the watt, the unit of electric power, and the kilowatt, which is a thousand watts. In metric shorthand, the watt is W; the kilowatt, kW.

As to that capital W, this is another case of a unit that is named after a person—James Watt.

If you have these units straight—the kilogram, the meter, the kilometer and centimeter and millimeter, the liter, the Celsius degree, and the watt and kilowatt—you're all set to step into the metric future.

## THE METRIC PAST

Early traditional measurements were based on things found in nature—barleycorns, the human foot, and such—resulting in a mathematical hodgepodge.

The idea of a simple, decimal measuring system started in France. About two centuries ago, the French Academy of Sciences decided to establish a unit of length based on a portion of the earth's surface. Surveyors spent seven years making measurements of a meridian—one of the imaginary lines running north and south from pole to pole. They determined, as accurately as they could, the length of one twenty-millionth of a meridian. Then scientists marked this exact length on a platinum bar. It became the standard for the new unit of length: the meter. Once the meter was set up, units of volume and weight were worked out from it. One tenth of a meter was cubed, and this became the liter. The weight of a liter of water became the kilogram.

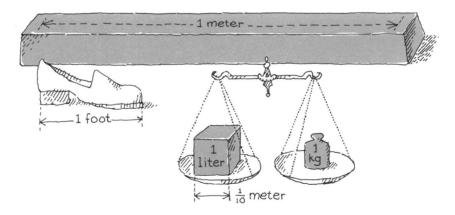

All neat and simple. So simple that many countries adopted the new system. In the United States, though, only scientists began using the new measures. They urged adoption of the metric system, and in 1886 a law was passed making the system legal.

Yet no law had ever been passed making the foot/pound system legal. Can you explain this?

... No law was needed to get people to use the customary measures. Feet and pounds had been used here ever since colonial times. They were old measures, brought here by English settlers.

The law about the metric system only said it *could* be used. It didn't say it *had* to be used. And most people didn't use it until recently.

### NOW THAT WE'RE GOING METRIC . . .

We're moving fast. Many industries and government agencies are adopting the new weights and measures. In many states, children are being taught *only* metric measurements. Feet and pounds are going the way of the horse and buggy. Yet most people do not know the correct name of the new system. It is officially called . . .

### ... LE SYSTÈME INTERNATIONAL D'UNITÉS.

The French name for the system is now being used throughout the world. Scientists everywhere have agreed to use this name only and not translate it into other languages. In English, we use only the initials SI.

SI is not quite the same as the original metric system. The old system has been incorporated into it and extended to include units for measuring electricity, temperature, and light. In addition, other units used by scientists are being redefined so that they fit in with the rest of the system.

Good-bye to all kinds of local measurements. All countries are beginning to use the new system. SI is truly international—the modern system for the modern world.

# Morton the moron starts his career

You know Morton. His sister, Cleo, thinks he is a moron. But she admits Morton is ambitious. He keeps wondering what he will be when he grows up.

One day, Morton was very quiet. Cleo found him in his room, cutting the buttons off his coat.

"Your good coat!" she screamed. "You're ruining it. Why, Morton . . . why?"

Of course, Morton still had to figure out how he was going to get rich. He soon had an idea. It came to him while watching traffic along the highway.

Cleo offered to help Morton. She made up a table explaining centimeters, meters, and kilometers. After Morton studied it, Cleo tested him.

Morton studied some more, but studying didn't seem to help. Cleo gave him other tests, and Morton failed them all.

But Morton did not give up. He went out and bought three metric measuring tapes. Cleo was puzzled when she saw them.

Morton, like all ambitious boys, wanted to work on Saturdays. He answered an ad in the paper, and Mr. Slink, the detective, hired him for odd jobs.

The first Saturday passed slowly. Mr. Slink had to be out of the office, but before he left he said, "Wash the windows and mop the floor. Don't let anyone in. And don't answer the phone. My answering service takes care of messages."

Morton soon forgot Mr. Slink's instructions. The phone rang and Morton picked up the receiver. Trying to sound grown-up, he said, "Slink Detective Agency."

The caller gave his name, but Morton had trouble understanding it. "What's your name?" he asked.

The answer sounded like "What." So Morton repeated, "What's your name?"

"Are you deaf? I've told you twice," said the man.

"Well, please spell it."

"W-a-t-t, Watt."

Morton hung up, muttering, "That guy is crazy. The inventor of the steam engine is dead."

The next Saturday, the job was more interesting. A bank had been robbed and Mr. Slink was called in to investigate. He took Morton to the bank with him, in case he needed an errand boy. On the way, Morton asked what had happened.

When they got to the bank, Mr. Slink started taking fingerprints.

Before Mr. Slink could explode, Morton changed the subject.

The phone was the cause of Morton's downfall. Morton kept on answering it when Mr. Slink was out.

One day, a lawyer had an important message for Mr. Slink.

When the lawyer told Mr. Slink the story, Mr. Slink wasn't amused—not one bit.

But Morton was not discouraged. He soon found another job. He became the weekend handyman for Farmer Brown.

Farmer Brown, as you may know, had several horses. One day, he was very upset because Frisky had run away.

Frisky had been trained to jump hurdles, and Farmer Brown was worried.

One day, Farmer Brown saw Morton nail some long, narrow strips of wood together. He watched Morton measure them.

Farmer Brown decided Morton was a moron. But he liked Morton. He really did.

And Morton liked him.

When Farmer Brown put Morton in charge of the vegetable garden, he tried to do a good job. He not only got rid of the weeds, but he got rid of the bugs.

# Square shooters

Here come some teasers with straight, true-blue answers.
But, careful now;
the questions might just be a little tricky.

**ROUND AND ROUND**

A long-playing record is 30 centimeters wide, and the label at the center is 10 centimeters wide. If the record has 70 grooves per centimeter, how far does the needle travel to play the whole record?

. . . **The needle travels 10 centimeters.**

How far the needle goes has nothing to do with the number of grooves per centimeter. Watch the needle on a record player and you will see that it doesn't go around and around. The record does.

## ANYTHING WRONG HERE?

"Good grief," said the photographer, "is the new Metric Queen a lady elephant? What measurements! Bust, 92; waist, 62; hips, 92. How can I make someone like that look good? I'll get pictures of a freezing fat lady. They say it's 28 degrees outside, and she has to be wearing a bathing suit for some of the poses."

"Calm down," said the designer. "In the next shots the model wears a size-38 evening gown and size-40 shoes."

"They'll never fit," said the photographer.

"Have you heard about the new metric sizes?" asked the designer.

The photographer groaned. "I've never used the metric system, and I never will—even if they do make it legal."

The designer walked away. She knew he was a good photographer, and she ignored the misstatements he had made. But can you spot them?

**. . . The Metric Queen is human and lovely,** for her measurements are in centimeters. Since the temperature is 28 degrees Celsius, she will be comfortable in her bathing suit. Her metric-sized dress and shoes will fit well.

Although the photographer doesn't know it, he uses metric units when buying photographic supplies. Film width is measured in millimeters. So are camera lenses. Besides, the metric system has been legal in the United States for over a hundred years.

P.S. One popular film width is 35 mm. Cameras using this size film are often called 35-millimeter cameras. The movies shown in theaters are usually projected from 35-mm film, but sometimes 70-mm film is used. This giant size is shown on wide screens and gives a very clear, bright picture. In most home movie cameras, the film width is only 8 mm—about as wide as a pencil.

## FAT MAN ON THE MOON

Have you ever seen movies of astronauts walking on the moon? They leap around like kangaroos even though they are wearing heavy space suits. They move easily because the moon's gravity is only one sixth of the earth's gravity, and they weigh only one sixth of their earthly weight.

Suppose that a fat astronaut tips the scales at 120 kilograms on the earth. Consider what would happen to him on the moon. How much weight would he lose? Would he still be fat?

. . . The astronaut would lose five sixths of his weight.

But he would still be fat. The amount of matter in his body—the mass—would stay the same. Wherever the astronaut is, his weight is due to the pull of gravity on his body mass. If gravity changes, so does his weight.

Everywhere on the earth's surface, the pull of gravity is just about the same. So we can say that a fat man on the earth weighs 120 kilograms when we really mean his mass is 120 kilograms. This habit of saying weight when we mean mass is very old, and nearly everyone does it. Even scientists often say weight instead of mass, and we do too in the rest of this book. As long as we stick to the surface of the earth, there is no confusion.

## WEIGHTY MATTERS

When is a hundred-dollar bill equal to a dollar?

... When you weigh them.

A hundred-dollar bill and a one-dollar bill weigh a gram each. With money, the most weight doesn't always mean the most value. A penny weighs three grams, but a dime weighs only two and a half. And a nickel just happens to weigh five grams.

You can use money as weights for a homemade balance scale. For the scale, you will need a round pencil, a rubber band, tape, two paper cups, a needle, some thread, and a ruler. Your old 12-inch ruler can be used. In fact, it's a good way to put it into service. This is what you do.

Twist the rubber band around the pencil several times. Then place the pencil across the center of the ruler. The flat side of the ruler should be touching the pencil. Bring the free end of the rubber band under the ruler, and wrap it around the pencil several more times. The rubber band should hold the ruler snugly underneath the pencil.

Next, hang a paper cup at each end of the ruler. Cut a piece of thread 50 cm long, and thread the needle with it.

Poke the needle through the cup at a point about 1 cm below the rim. Then run the needle through the opposite side of the cup. Remove the needle and tie the ends of the thread together. Repeat this for the second cup and you have a loop of thread running through each cup. Tape one loop to one end of the ruler and the other loop to the other end.

Now rest the ends of the pencil on two chairs or two tall piles of books that are the same height. Adjust the ruler until it balances, and you are ready for weighing.

Place a small object such as a key in one of the cups. Put money weights into the other cup. Try different combinations of pennies, nickels, dimes, and tightly folded dollar bills until the scale balances. Empty the cups and add up the gram value of the weights. A dollar is 1, a dime is 2.5, a penny is 3, and a nickel is 5. The total is the weight of the object in grams.

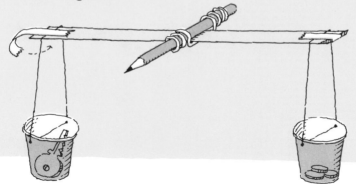

Now maybe you can answer a very weighty question. What can't you hold for ten minutes even though it weighs less than a gram?

. . . **The air in your lungs.** Time yourself and see how long you can hold your breath.

## OLDEN AND GOLDEN

Do you know the oldie: Which is heavier—a pound of feathers or a pound of gold? A pound is a pound, isn't it? Well, not always—and that's the catch. An avoirdupois pound is heavier than a troy pound. Both are units in the customary system. Which one is used depends on what is to be weighed. For feathers, it's avoirdupois pounds; for gold, it's troy pounds. So a pound of gold is lighter than a pound of feathers.

Now can you say which is heavier—a liter of milk or a liter of cream?

. . . **The liter of milk is heavier.**

There is only one kind of liter, and it's a measure of volume—not weight. Since cream floats on top of milk, the milk must be denser, so a liter of milk is heavier than a liter of cream.

## THE RAT TAKES THE CHEESE

In playing farmer in the dell, the cat takes the rat and the rat takes the cheese. That's how the game has always been played, but this time the rat is hungry and there are two pieces of cheese. One piece is a bar 3 cm × 4 cm × 10 cm, and the other is a chunk, 120 mL. Which one should the rat take to get the bigger piece?

. . . It doesn't matter which one is chosen.

A cubic centimeter ($cm^3$) equals a milliliter (mL), which is one thousandth of a liter. The volume in the bar equals 120 $cm^3$. So there's exactly the same amount of cheese in the bar as in the 120 mL chunk. No more, no less.

## HELIUM FOOTBALL

During a football game, a coach suspected the opposing team of cheating. They were kicking very well. Too well, it seemed. Why were the kicks staying in the air longer than usual, the coach wondered. It occurred to him helium is much lighter than air. Could it be that the football was filled with helium?

. . . Maybe it was and maybe it wasn't.

The air in a standard football weighs about 18 grams; the helium filling the same size ball at the same pressure would weigh about 3 grams—a difference of 15 grams. This difference is too small to matter, for the football itself weighs about 400 grams.

## APPLES AWAY AND AWEIGH

When will apples be sold by mass instead of weight?

**. . . When people colonize the moon and other low-gravity places.** Probably food will be grown here and dried. Then it will be sold by mass before it is shipped off into space.

### CAR PROBLEMS

In the United States, speed has been measured in miles per hour. But now some highway signs give speed limits in kilometers per hour. And every car made in the United States after 1976 has a metric scale on its speedometer. Eventually, highway signs will be completely metric, and all cars will have to have metric speedometers.

Suppose when that happens you inherit a working antique car. But how can you drive safely with an out-of-date speedometer? Keep the car. You can fix the speedometer in minutes.

Modernize the speedometer. Here's how.

Take some heavy paper, a pencil, a pair of scissors, and some tape out to the car. Cut a doughnut-shaped piece of paper to fit around the speedometer gauge. Tape it in place.

Mark a kilometers-per-hour scale on the paper. Ten kilometers is 6.2 miles. Between six and seven, but nearer to six, make a mark. Write 10 next to the mark.

Twenty kilometers is 12.4 miles. About half-way between twelve and thirteen, make the 20 kilometer mark.

Keep on in this way. Work around the gauge and mark off every ten kilometers.

This chart will help you in placing the marks correctly.

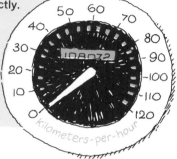

30 km = 18.6 miles
40 km = 24.9 miles
50 km = 31.0 miles
60 km = 37.3 miles
70 km = 43.5 miles

80 km = 49.7 miles
90 km = 55.9 miles
100 km = 62.1 miles
110 km = 68.3 miles
120 km = 74.6 miles

After marking 120 km, just write: Whoops! Too fast!

## QUESTIONS FOR BACKSEAT DRIVERS

1. The distance from Main Street to the stop sign at the intersection with the highway is 1½ kilometers. One day, a driver started at Main Street and, after going one kilometer, he reached a steady speed of 20 kilometers per hour. He maintained this speed for four full minutes, covering a total distance of 2⅓ kilometers. At no time did the driver slow down. Did he illegally pass the stop sign without stopping?

2. What part of the road gets longer when something is taken away from it?

3. How can a car keep up with a jet plane that is traveling at 900 kilometers an hour?

55

1. No, the driver turned around and drove away from the highway.
2. A drainage ditch under construction along the side of the road.
3. The car is in the cargo hold of the jet plane.

## LION ON THE MOVE

The circus was leaving town, and the movers had to get a caged lion into a waiting train. The cage stood twenty meters from the train and was much too heavy to carry. The movers managed to lift the lion cage onto three logs to be used as rollers. They had no trouble pushing the cage on the rolling logs. Every time the back log rolled out from under the cage, the movers put it in front and rolled the cage onto it.

Each log measured one meter around, and the movers finished the job quickly. Why were they so fast? Well, figure out how many times the logs turned around in moving the cage twenty meters.

... Ten is the right answer.

Surprised? As the logs rolled around once, they moved forward one meter. But the cage also rolled forward one meter on top of the logs. So for each complete turn they made, the cage moved forward two meters.

Still not convinced? Then, try an experiment. Put a cereal box on some paper rollers. See how far it moves when the rollers make one complete turn.

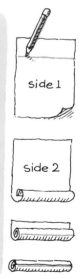

To make a roller, take a sheet of typing or notebook paper about 22 cm by 28 cm. Mark each side 5 cm from the top. Using the marks as guides, rule a line across the paper. Place the paper on a table with the marked side down. Starting at the edge farthest from the mark, roll the paper tightly. As you finish rolling, loosen or tighten the paper so the top edge meets the ruled line. Tape the paper so that the roller holds its shape. To make sure it will be smooth, run the tape along the entire line. This will give you a roller 5 cm around.

Make a second roller like the first. Since you are going to move the box only a short distance, you need only two rollers.

To move the box, first put the rollers on the floor. Make sure the ruled lines are on top. Place the box on the rollers so that about a third of it sticks out in front and back of them. Mark the starting position of the box. Now hold the box and shove it gently until the rollers have gone around once.

Measure how far the box has moved. Is your answer 10 cm? It should be if the rollers are 5 cm around.

## HARDWARE AND SOFTWARE

Now that the United States is going metric, business people often speak of hardware and software changes. Their talk is not easy to follow, for hardware is not always hard, and software is not always soft.

Any piece of standard equipment—a wrench, a machine, a computer, a thermometer, or even a flexible measuring tape—is hardware. Software, on the other hand, is printed matter: documents, records, computer programs, and labels.

It doesn't cost very much to introduce the metric system in software, and the changeover is going on very rapidly. The changeover in hardware is another matter. Many industries will have to spend enormous amounts of money to go metric, and so they will continue to use old measures and measuring units for many years. It's not that kilogram scales and centimeter tapes are so expensive. New metric instruments can be low in price and still be costly.

How do you explain this?

. . . The problem is not the measures—but what they measure.

For instance, every nut and bolt in an American car is measured in inches. The standard metric parts can't hold one of these cars together, because all of them are a bit too small or too large.

New devices for measuring nuts and bolts and other parts do not cost a lot, but making the parts in new sizes means expensive changes in machinery. Changes in an entire assembly line may be needed, and this may cost millions and millions of dollars.

The next time you go shopping, see how many changes in software have already taken place. Sheet sizes, for example, used to be given in inches. Now they are being marked in both inches and centimeters. Soon, only centimeters will be used. The sheets won't be any different, but the labels will.

Take a look around your kitchen and see how many food packages give the contents in metric quantities. Can you find some labeled with both customary and metric units? Notice pudding, cocoa, soup, and cereal labels. Most brands now give quantities in both units, but this practice won't last long. It's really a good idea to collect double-unit labels. They will be valuable antiques someday!

You can expect changes in the labels on almost everything that's sold in supermarkets in the near future.

But there is one item that probably won't change. It is something you cut and pass around at the table but never eat. Any idea what it is?

. . . A deck of cards.

## BOTTLE BATTLE

The title sounds like a slapstick comedy, but it has nothing to do with old movies. Bottle battles are going on right now—in supermarkets.

Many different bottle shapes are appearing on store shelves, fighting for the attention of customers. Notice that salad oil, liquid detergent, and shampoo are often sold in hourglass-shaped bottles and bottles with handles. Many companies prefer them to the ordinary kind, with straight sides.

Do you know why?

. . . Money, money, money. Companies usually make more from products in dented bottles. They sell faster.

A dented bottle is taller and looks bigger than an undented bottle holding the same amount. People think it is the better buy, although usually it isn't.

Looks can be deceiving. Reading the label is the best way to tell how much a bottle holds.

**SHOPPERS' HEADACHES**

Good-bye, gallons. Hello, liters—three liters. In the metric changeover, soda pop and wine lead the way. In place of gallons, drinks are being bottled in three-liter jugs, which are as tall as—or taller than—the old gallon containers.

Many leading brands of soda are now sold in a one-and-a-half (1½) -liter bottle instead of the old, half-gallon bottle. In fact, the half-gallon bottle might be another item to add to your collection of future antiques.

Meanwhile the new containers are causing a lot of confusion. Here are some of the problems that shoppers face:

1. Suppose LOAD-A-SODA now comes in 1½-liter bottles that sell for 65¢. It used to come in half-gallon bottles selling for 69¢. The new bottle seems to be the better buy. Is it?

2. If a cubic centimeter equals a milliliter and there are one thousand milliliters in a liter, how many milliliters of milk are there in a one-liter container?

3. A container that measures 5 cm × 10 cm × 20 cm can hold a liter of popcorn. If all the dimensions were doubled, how much popcorn would the new box hold?

1. **The half-gallon bottle is the better buy.**
   A half gallon equals 1.892 liters. That's .392 liter more than 1.5 liters. The old bottle holds nearly two glasses more than the new one.
   Want to buy stock in the LOAD-A-SODA company?

2. **That depends on how much milk was put into the container.**
   Maybe it wasn't filled.

3. **Try again if you said twice as much.** The new box would contain 10 cm × 20 cm × 40 cm. That comes to 8,000 cm³, or 8 liters. So it would hold eight times as much.

## CELSIUS SCALE TUNE-UP

1. The average temperature in Detroit in February is 26 degrees. About ten kilometers southeast of Detroit, the temperature averages three degrees below zero during the same month. How is that possible?

2. Ice freezes at zero Celsius (0°C). Does ice ever get colder than that?

3. "What strange weather we're having!" said Sam to the weatherman. "It's May and it's still chilly. Can you tell me how cold it's been during the past four days?"

   "I don't have the figures at hand, but I do remember that the highest temperature each day was one degree higher than the day before. I also remember that the high points added up to 40°C, although they averaged 10°C."

   Sam figured out the temperatures. Can you?

1. **The border between the United States and Canada is not a straight line.** Look at a map and you will see that Windsor, Canada, is southeast of Detroit, U.S.A. The Canadians have gone metric, so the temperature at Windsor is measured in Celsius degrees. And minus three degrees Celsius (−3°C) is the same as 26 degrees on the Fahrenheit scale.

2. **Ice can be colder than the freezing point.**
   Have you ever touched an ice cube and felt your finger stick to it? That ice cube was so cold it froze the moisture on your finger. It's temperature was below zero, the freezing point on the Celsius scale.

3. **The four high points were: 8.5, 9.5, 10.5, and 11.5 degrees Celsius.**

## MYSTERY MONEY

Each year, the owner of a tool company bought 10,000 kilograms of steel. The steel was made into tools, which were sold at a profit. Usually the owner made $250,000. This went on year after year, but neither the company nor the owner ever paid the United States Government one penny of income tax.

How did they get away with that?

*. . . The owner, Mary Burns, was a Canadian citizen living in Canada and running a Canadian company.*

### METRIC FINE POINT

The metric system was designed to be logical and easy to use, but it still has some rough spots. Take highway speed, for example. On many roads in the United States, signs now give the limit as 90 kilometers an hour. But that's not a metric rate. It's only partly metric. Can you explain this?

*. . . Hours are not metric units.* In the metric system, the second is the only unit of time. If it were used, highway speed rates would be in meters per second. That would be very inconvenient for drivers, and so all metric countries have agreed to use kilometers per hour.

### THE LITER THAT WASN'T A LITER

In setting up the metric system, the liter was defined as the volume of a special amount of water. One liter of water was supposed to equal the weight of the kilogram standard kept at the International Standards Organization, in France. Experts thought the amount of water with exactly the same weight as the kilogram standard took up exactly 1,000 cubic centimeters. As it happened, they were not exactly right.

Years later, when metric standards were checked with more accurate instruments, the liter was found to be a pinhead larger than 1,000 cubic centimeters. It was 28/1000 of a cubic centimeter more. This tiny mistake created a big problem, for all metric units are related to each other. Many scientists began to avoid using the liter even though it was a convenient size to work with. The mistake had to be corrected, and it was. Metric experts soon found a way to solve the problem.

Can you figure out how they did it?

... The definition of the liter was changed.

Experts decided not to use the volume of a kilogram of water in determining the size of the liter. There was no problem with the cubic centimeter, and so the liter was defined as 1,000 cubic centimeters. That's how much space it was supposed to take up in the first place.

## STILL TO COME

Just about every aspect of your life is covered by SI [Système International]. All the substances you take into your body and all the waste you give off can be measured metrically. There are SI units for electricity, heat, and the other kinds of energy you use and produce. But there is one thing you do that can't be measured with an SI unit at present. All you have to do is . . .

. . . yell.

The unit for measuring the loudness of sound that's now in use is the decibel. But it's not related to the metric system.

One of these years, this problem will be taken care of by the international officials who are working on making SI the best and most comprehensive system of measurements the world has ever known.

# Klondike Mike and the solid that could be poured

Gold rush days, 1896. Frenchy DuBois, from Paris, and his son, Michel, had reached the Klondike. Mike expected to help his father pan for gold. But the only mining he ever did was indoors, inside the Black Horse Saloon.

Mike had discovered how to separate two solids that can be poured. These were . . .

... sawdust and gold.

This is how it happened. DuBois decided to make the first prospecting trip alone, while Mike stayed behind, working at the Black Horse Saloon.

Mike had no time to miss his father. He cooked, washed dishes, and cleaned up the barroom. He swept up the sawdust every night, then covered the floor with a fresh batch.

Sandy Simon, the saloon owner, liked Mike, but he thought the boy's habits were strange. He saved all the sweepings, and when he went to bed, he carried them to his room, along with a bucket of water.

Sandy was also bothered by Mike's interest in the gold scale. After all, every saloon had one, for miners often paid their bills with nuggets or gold dust.

Sandy became suspicious when Mike asked to take the scale and weights to his room overnight. But he let the boy borrow the set. Why did he want it, Sandy wondered.

Was Mike stealing gold?

Mike was secretive. But he was not a thief. He was mining gold from the barroom floor. A few grains were always in the sawdust. They came from mud on the miners' boots and from gold dust that spilled when men paid for drinks.

Collecting gold was easy. Mike put the sweepings in water, and while the sawdust floated on top, the gold sank.

After a month, Mike had almost an ounce. Any day, now, his father would be back and see his treasure.

But his father never returned. A miner reported that he had found gold, and then bandits had robbed and killed him.

Sandy tried to console Mike.

Mike stayed until he was sixteen. By then he had a small fortune in gold dust. He packed it in little bags and sewed the bags in the lining of his coat.

Some men were leaving for San Francisco, and Mike joined them. After a trek overland, a trip by train, and another by steamer, he finally reached the United States.

Mike was on his own now. He went to California's famous Palace Hotel and tried to get a room.

News of his gold spread fast. Strangers asked him for loans, swindlers offered him deals, and city toughs tried to rob him.

Mike decided to move on. He packed his bag, bought some maps and a horse, and headed for the Rockies.

But the bandits along the trail were worse than city holdup men.

Mike was in mountain country when a gang spotted him. They chased him down a ravine. But Mike was lucky. He found a cave and hid there for almost a week.

Before leaving, he poured his gold dust into two strong sacks. He buried one in the cave and the other in an abandoned mine a few kilometers away.

Mike did not trust his memory, but he trusted the robbers even less. He made a map showing where the gold was hidden, and he wrote some notes in code. If robbers ever caught him and found the notes, they wouldn't know what they meant.

"The gold is safe," Mike thought as he hit the trail again. He kept riding until he came to a train. It was eastbound, and Mike boarded it.

Mike settled in Chicago. As the years passed, he had a son, who had a daughter, who had a son who was called Little Mike.

Klondike Mike was over eighty now and his great-grandson adored him. Little Mike never had enough of the stories about the bad old days. But the boy thought they were just stories. His favorite was the one about the hidden gold.

That never happened. Klondike Mike died, and Little Mike grew up.

He was in college when he decided to go hiking in the West, carrying his great-grandfather's pack. Mike emptied the pack, and a pair of old overalls fell out. In one of the pockets he found this strange note:

The Navajo call it upper because of the 3000m high mountains.

It is 750 km east of the great ocean.

You can find it by seeking **the four** in the middle of this message...

CARE! YOU MAY GO, BUT AHEAD LIES DANGER.

Four what? I GOT IT...

... the four letters spell UTAH.

Klondike Mike had often told stories about his travels in Utah. And he had said the state name was an Indian word. He even said it meant *high*.

Klondike Mike had gone to school in France and learned the metric system there. He knew m stood for meter; km, for kilometer. Here was the key to the secret code!

In another pocket, Mike found another note like this:

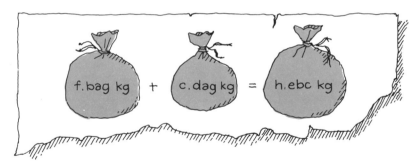

No doubt about it . . . kg meant kilogram, and the bags were the sacks Klondike Mike had filled with gold. The rest wasn't hard to figure out, for the + and = signs were clues. Probably the dots were decimal points and the letters were coded numbers. Reasoning that *a* stood for 0, *b* for 1 and so on, Mike replaced the letters with numbers.

This is what he got: $5.106 \text{ kg} + 2.306 \text{ kg} = 7.412 \text{ kg}$

Was there really all that gold in the bags? Seven kilograms would be worth thousands! But where in Utah was it? WHERE?

There must be a map, somewhere. Mike ripped the overalls apart, to look for secret pockets, and in one of the cuffs, he found this paper:

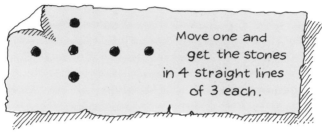

Move one and get the stones in 4 straight lines of 3 each.

No, that wasn't a map. Mike nervously opened the other cuff. A neatly folded sheet fell out. Sure enough, it was a map—THE map!

The map showed the Rocky Mountains in eastern Utah. Only one town was on it: Lonelyville.

On the back were these directions in faded writing:

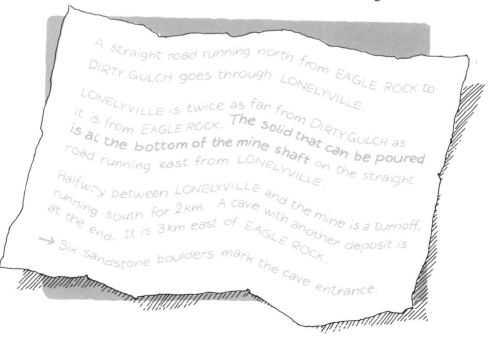

A straight road running north from EAGLE ROCK to DIRTY GULCH goes through LONELYVILLE.

LONELYVILLE is twice as far from DIRTY GULCH as it is from EAGLE ROCK. The solid that can be poured is at the bottom of the mine shaft on the straight road running east from LONELYVILLE.

Halfway between LONELYVILLE and the mine is a turnoff, running south for 2km. A cave with another deposit is at the end. It is 3km east of EAGLE ROCK.

→ Six sandstone boulders mark the cave entrance.

Mike was so excited, he had trouble understanding the directions. He tried making a map from them, and this was the result:

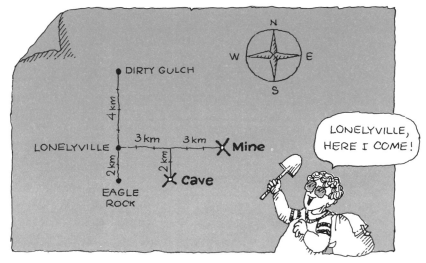

Mike arrived by plane. At the airport, he saw an old man and asked him for directions.

Mike pulled out his pedometer to measure the distance and started walking.

He walked 6 km on Highway 67 and there was the garage. What had happened to the gold, Mike wondered.

Mike backtracked on Highway 67, hoping to find the cave. Only part of the cave remained. Most of it had been blasted away to make room for a Burger Queen restaurant, but the deepest part was still there.

Off in the distance were the six stone boulders shown on the old diagram. They were still in the same positions.

Klondike Mike had said move one to make 4 rows of three. But which one? How? Then suddenly Mike caught on. . .

. . . If one rock was moved to the spot on which he was standing, there would be 4 rows. That meant the hidden gold was nearby!

Mike started digging, but he didn't get very far.

Mike was taken to court. When he tried to explain what he was doing, the judge only said, "Thirty days. The jail is across from the Burger Queen. You can look for gold there."

And Mike did. His job was to clean the main hall of the jail. Men often came in with boots covered with glittering mud. The flecks in it probably were Klondike gold dust from a buried sack that had broken. Mike began to collect the sweepings. And just as his great-grandfather had done, he started mining the solid that could be poured. . . .

# When you're looking for something to do

.... and it's raining outside and you're sick of watching television, cheer up. Here are lots of tricky things you can do and make, using your metric know-how.

As a starter, how about making a key that balances on your finger? Here are the directions for this unusual gadget. . . .

## THE BALANCING M

The M stands for meter, and you can balance it on your finger. It really is a key—one that explains everyday metric measures.

To make the M, you need: a pencil, thin paper, scissors, and a piece of lightweight cardboard.

First, make a pattern. Place the thin paper over the M shown here. Trace it and cut out the letter. Then trace your pattern on the cardboard. Cut it out and you have an M that will balance on your finger. Test it.

Here is the information to go on one side of the M:

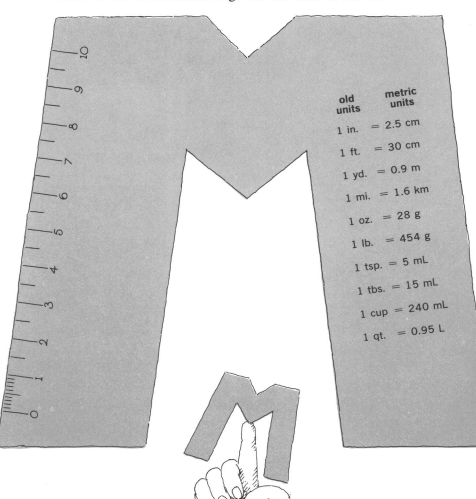

| old units | metric units |
|---|---|
| 1 in. | = 2.5 cm |
| 1 ft. | = 30 cm |
| 1 yd. | = 0.9 m |
| 1 mi. | = 1.6 km |
| 1 oz. | = 28 g |
| 1 lb. | = 454 g |
| 1 tsp. | = 5 mL |
| 1 tbs. | = 15 mL |
| 1 cup | = 240 mL |
| 1 qt. | = 0.95 L |

To put the key on the M, start by marking off a 10-cm ruler along the edge of one leg. On the other leg, write the table that explains how to change the old units into metric units. All the main measurements will fit if you use abbreviations and symbols, as in the picture.

On the other side of the M, write the table to use in changing metric units into customary units. Put it all on one leg. On the other leg, put a table for oven temperatures in Celsius and Fahrenheit.

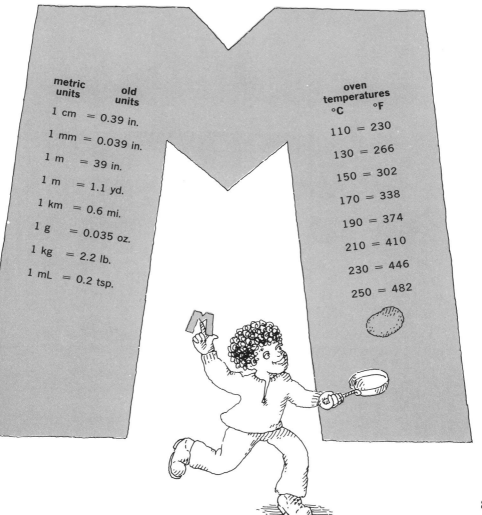

| metric units | old units |
|---|---|
| 1 cm | = 0.39 in. |
| 1 mm | = 0.039 in. |
| 1 m | = 39 in. |
| 1 m | = 1.1 yd. |
| 1 km | = 0.6 mi. |
| 1 g | = 0.035 oz. |
| 1 kg | = 2.2 lb. |
| 1 mL | = 0.2 tsp. |

| oven temperatures | |
|---|---|
| °C | °F |
| 110 | = 230 |
| 130 | = 266 |
| 150 | = 302 |
| 170 | = 338 |
| 190 | = 374 |
| 210 | = 410 |
| 230 | = 446 |
| 250 | = 482 |

## KITCHEN UPDATE

Like to cook? Then, become a metric chef. Try your hand at some recipes that give quantities in metric units. You'll need metric measures for them, of course, but that's no problem. You can make the measures yourself.

For small quantities, you will want milliliter measures. Since a teaspoon equals 5 mL and a tablespoon equals 15 mL, you can use a set of regular measuring spoons. Just mark them correctly.

For larger quantities, you will want a measure that holds 250 mL. As you know, 250 mL equals 250 cm$^3$, or a quarter of a liter. It's easy to make a measure of this size from a glass jar if you follow these directions.

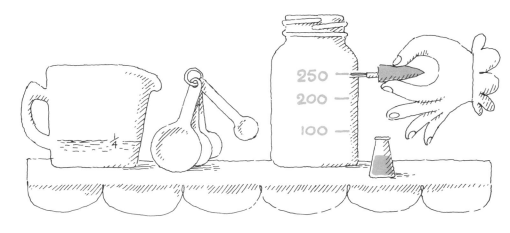

First collect the materials you need: a medium-size jar, some nail polish, an old-style measuring cup, and a set of measuring spoons.

Pour water into the measuring cup until it reaches the ¼ mark. Put this water into the jar. Next pour 3 tablespoons of water into the jar. Make sure the tablespoon is full to the brim each time you use it. Now dip out ½ teaspoon of water. You have 100 mL of water in the jar. Mark the level of the water with a thin line of nail polish.

To make the 200-mL mark, add another ¼ cup plus 3 tablespoons of water. Dip out ½ teaspoon of water, and mark the level with nail polish.

The last measurement will give you 250 mL. Add 3 tablespoons plus 1 teaspoon of water, and mark the level.

Ready to try out the new quarter-liter measure? If you want to make something that's really good—but simple—line up the ingredients for:

### Beautiful Berry Milk Shake

**200 mL fresh berries**
**500 mL milk**
**25 mL sugar**

Any kind of berry can be used. Blueberries that you have picked yourself are marvelous. Strawberries are good too. If you use them, cut them into small pieces.

Put the milk, berries, and sugar in a mixing bowl. Beat for a minute with an electric mixer or an eggbeater. Then pour into glasses and serve.

## WHEN YOU COOK DINNER

Here's a delicious meal for six to eight people. It is made of two layers baked in the same pan. The recipe calls for a square baking pan 23 cm on each side, but any pan will do if the bottom area is about 530 cm². 

Preheat the oven to 200°C. Then get together the ingredients:

### Corn-bread-and-bean Double Feature

**THE BOTTOM LAYER**

100 mL margarine
2 large onions
2 454-g cans kidney beans
2 227-g cans tomato sauce
10 mL chili powder

**THE TOP LAYER**

250 mL yellow cornmeal
250 mL flour
50 mL sugar
15 mL baking powder
5 mL salt
1 beaten egg
75 mL margarine
250 mL milk

Cut the onions into very small pieces. Fry them in margarine until you can almost see through them. Add the kidney beans and liquid from the can, the tomato sauce, and the chili powder. Cook for 5 minutes, stirring often. Then set aside.

Mix the dry ingredients in a large bowl. Add the rest and beat the mixture with an eggbeater or an electric mixer until it is smooth.

**THE COMBINATION**

Spread the bean mixture in the baking pan and cover it with the cornmeal batter.

Bake for ½ hour at 200°C, until the top is golden brown. Let it cool 5 minutes before serving.

## GREAT ANY TIME

### Marvelous Metric Pancakes

300 mL flour
5 mL salt
45 mL sugar
10 mL baking powder
2 eggs
45 mL oil
200 mL milk

Mix the ingredients together in a large bowl. Beat them lightly with a fork.

Get the griddle ready by rubbing it with a thin coat of oil. Heat the pan, then test it with a drop of cold water. If the drop sits in one spot and boils away, the griddle isn't hot enough. If you have the right temperature, the drop bounces around when it lands.

Pour the batter from a ladle or a large spoon held near the griddle. The pancakes should be about 10 cm wide. Watch for bubbles. When they form, turn over the pancakes and cook the other side. This recipe makes about fourteen pancakes—enough for three people who are quite hungry, or for two who are starving.

### GRIDDLE DIDDLE

When you tested the pancake griddle to see if it was hot enough, you dropped cold water on it. Did you notice that the water drops lasted longer when the griddle got hotter? When it is fairly hot (180°C), the drops boil away quickly. When it is very hot (230°C), the water drops last longer before they sizzle away. Can you figure out this mystery?

. . . As a water drop gets near the very hot griddle, the bottom layer turns to vapor. The vapor is a poor conductor of heat and it keeps the drop from heating up instantly. If the griddle is only fairly hot (180°C), the drop doesn't vaporize when it lands. Instead, it wets the pan and boils away faster.

## DOUGHY CREATIONS

Everyone eats bread and cake. But why not be different and *wear* something made of dough? Here's a mixture that can be used in making jewelry and gadgets for yourself and friends.

### Metric Modeling Dough

100 mL cornstarch
200 mL baking soda
125 mL water

Put the cornstarch and baking soda in a saucepan and mix them thoroughly. Stir in the water.

Heat the mixture carefully, stirring constantly with a large spoon. Cook the mixture until it becomes so stiff that when you stick in a spoon, the spoon stands up. Remove from heat and continue stirring for 30 seconds.

Press the dough together with the spoon and cover it tightly with aluminum foil or a damp cloth. Allow the dough to cool for at least two hours. Knead it a few times and it's ready to use. The recipe is enough for about ten creations.

If some dough is left, wrap it tightly in aluminum foil and store it in the refrigerator. It will keep for two weeks or so.

The first time you use the dough, try some of the pieces shown here. After that, let your imagination go.

### Millimeter Jewelry

Copy the design you plan to use onto a piece of plain white paper. Then put a ball of dough on it. Pat and mold the dough until it fills the outline. A large shape should be about 8 mm thick, but 5 mm is enough for the raindrops.

Make holes with a toothpick or a needle for a neck string. Place the holes so that, when the string goes through them, the jewelry will lie flat.

Allow the jewelry to dry overnight. Do not handle it while it is drying, for damp dough is crumbly. After it is completely dry, it is very strong.

Use watercolor to paint the dry jewelry. Then coat it with colorless nail polish. Put string through the holes and show off your handiwork.

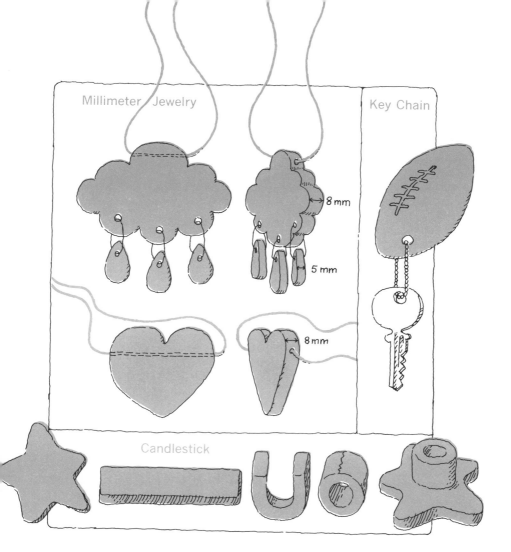

### Key Chain

The football should be about 5 cm long and 2.5 cm thick. Make a hole wide enough for a chain. Allow the creation to dry. Then paint and coat it with nail polish. Add a chain and your keys.

### Candlestick

The candlestick is made by joining a ring of dough to a flat star. Follow the pattern to make the star. Make the ring from a rectangle 2.5 cm by 12 cm. Each should be 8 mm thick.

Join the pieces securely. Smooth the seam between them until the candlestick looks as if it is made of one continuous piece of dough.

Dry, paint, and polish it. Then stick in a candle. Pretty, isn't it?

## UPSIDE DOWNERS

Each of these riddles has a one-word answer. Can you guess this word?

To check your guess, use a pocket calculator. Add up all the figures given in the riddle, unless you are told to multiply or subtract them. Notice the result in the window. When the calculator is right side up, it reads as a number. But when the calculator is turned upside down, the digits look like letters. Read them that way, and they spell the word answering the riddle.

If you don't want to bother turning the calculator upside down, skip that step. Just stand on your head and read the answer.

1. What do you call someone who is 151 cm tall and weighs 234 kg (multiply) and eats 46 cupcakes a day?

2. What weighs 524 g, is 28 cm long × 10 cm wide × 9 cm high (multiply), and holds 1 solid foot?

3. The load is 156 cm high and weighs 48 kg, and it takes you 2 hours to wheel it 7500 m in a wheelbarrow. Add to this the fact that a strong young fellow named Bill can't even wheel it 10 m in 2 hours. What is in the wheelbarrow?

4. A tank weighing 11,500 g held 42,000 g of water. It was full. Yet some items, each less than a hundred square millimeters (mm²), were put in the tank. They measured 88 mm², 31 mm², 63 mm², and 22 mm². What were the added items?

Were these your answers?

1. OBESE (151 × 234 + 46 = 35380)
2. SHOE (28 × 10 × 9 + 524 + 1 = 3045)
3. BILL (156 + 48 + 2 + 7500 + 10 + 2 = 7718)
4. HOLES (11500 + 42000 + 88 + 31 + 63 + 22 = 53704)

### DOWNSIDE UPPERS

These are reverse riddles to be solved with a calculator. But you don't start with the question. Instead, you start with the answer: the word that appears in the window when the calculator is held upside down. Then you make up a calculator riddle that is answered by that word.

This type of riddle is quite tricky. The first step is to write the word backwards. Then you substitute digits for the letters, using this key:

0 = O   1 = I   3 = E   4 = H   5 = S   7 = L   8 = B

Try starting with HOSE. That gives ESOH, and that gives 3504, which might be the answer to this riddle:

What things are in a box 25 cm long, 25 cm wide, and 16 cm high, that can be laid end to end so that they come to 3438 cm?

... 25 + 25 + 16 + 3438 = 3504, and the hose are nylons.

Now try working out a calculator riddle answered by the word EELS. You could ask . . .

. . . A fisherman's wife, working in a store in Venice, Italy, was 1871 mm tall and her arm span was also 1871 mm from the tip of the middle finger on her left hand to the tip of the middle finger on her right hand. Her bust measured 971 mm and her hips, 1020 mm. What did she weigh?

. . . Since 1871 + 1871 + 971 + 1020 = 5733, it is clear that she weighed EELS. After all, eels are a popular food in Italy.

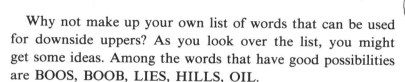

Why not make up your own list of words that can be used for downside uppers? As you look over the list, you might get some ideas. Among the words that have good possibilities are BOOS, BOOB, LIES, HILLS, OIL.

Still raining? Are you in a mood for something different? Then try these:

### ZANY DAY TEASERS

1. Every word in this book can be put on one *edge* of a sheet of typing paper 21 cm × 28 cm. Can you figure out how to do this?
2. Your head probably measures more than 45 cm around; a little-finger ring measures 5 cm around. How can you push your head through such a ring? (If you don't have a small ring, take a strip of paper 5 cm long and tape the ends together.)
3. How can you balance a one-liter bottle on a tightly stretched rope suspended in midair? (Hint: Try using an umbrella.)

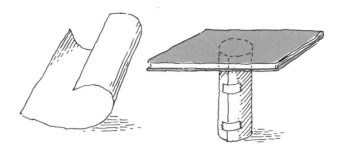

1. **Roll the paper into a cylinder about 5 cm across.** Tape the edge so that cylinder stays rolled. Stand the cylinder on a table. Now put the book on the edge of the rolled paper. Balance the book and it will stay there, supported by the cylinder.

2. **Hold the ring near your head.** Put your little finger through the ring and push your great big head.

3. **Like this:**

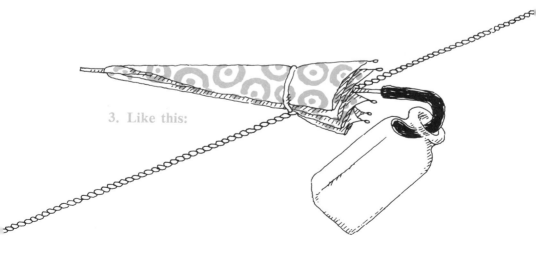

## CUTUP

Can you fit these pieces together so that they make a surprise scene? Trace them and cut them out.

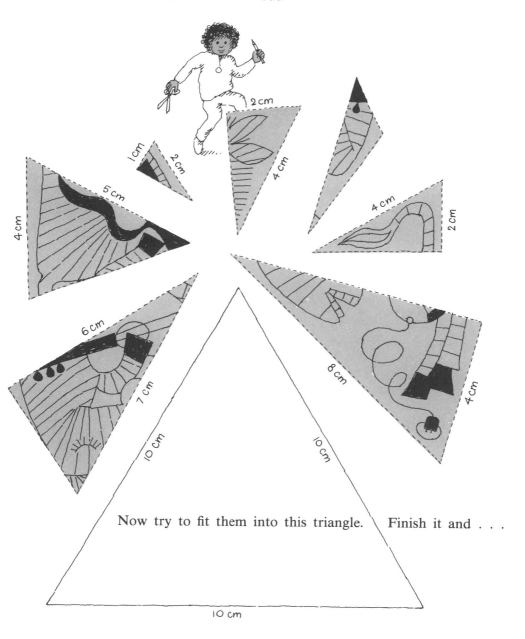

Now try to fit them into this triangle. Finish it and . . .

. . . the scene looks like this:

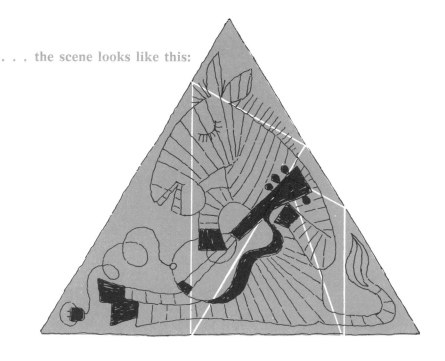

## WHATEVER THE WEATHER

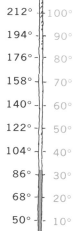

Whether the weather be cold,
Or whether the weather be hot,
We'll weather the weather,
Whatever the weather,
Whether we like it or not.

So goes an old English song from the days before people had comfortable outdoor clothing.

But what do you wear when the temperature is 32°C? A ski suit or a bathing suit? Plan to swim. Thirty-two degrees Celsius is hot.

Confused? Your old Fahrenheit thermometer won't help. But you can change it into a Celsius thermometer by putting a new scale on it.

> Cut a strip of paper to fit along the old scale. Tape it in place. Then put Celsius marks that match the Fahrenheit scale on the strip. In matching them, use the table on the left.

## WIND WATCHING

Although you can't see the wind, you can observe its speed—and without instruments. Just use your eyes and this chart.

| Speed kilometers per hour | Name of the wind | Observing the wind's work |
|---|---|---|
| less than 2 km/h | calm | smoke goes straight up |
| 2 to 5 km/h | light air | smoke is bent |
| 6 to 11 km/h | light breeze | leaves rustle; weather vane moves |
| 12 to 19 km/h | gentle breeze | leaves and twigs in constant motion |
| 20 to 29 km/h | moderate breeze | raises dust, moves small branches |
| 30 to 39 km/h | fresh breeze | small trees sway |
| 40 to 49 km/h | strong breeze | big branches in motion |
| 50 to 61 km/h | moderate gale | whole trees sway |
| 62 to 74 km/h | fresh gale | breaks twigs off trees |
| 75 to 87 km/h | strong gale | damages chimneys and roofs |
| 87 to 101 km/h | whole gale | trees uprooted |
| 102 to 120 km/h | storm | widespread damage (very rare inland) |
| above 120 km/h | hurricane | most destructive of all winds |

Check your observations with the Weather Bureau reports. The reports from Canada are all metric now, and those from the United States will soon be metric too.

## IT'S RAINING—IT'S POURING

And the day need not be boring. While you wait for the weather to change, measure the raindrops.

The largest raindrops—the kind that patter and splatter as they land—are rarely more than 5 mm wide. The smallest are about 0.5 mm wide. Drops below that size are called drizzle.

You can get some idea of the size of raindrops in this way:

First take an eyedropper and measure its width. This gives the size of the drops that fall from it.

Fill the eyedropper with water and hold it high above a pan of flour while you squeeze out a few drops. Notice the doughballs that form.

Cover the doughballs with aluminum foil. Then set the pan outdoors for a moment while raindrops splash on the flour.

Take the pan indoors, uncover it, and compare the doughballs from measured drops with those made by raindrops.

## METRIC RAIN GAGE

You might also want to measure the amount of rainfall during a storm. To do this, you use a rain gage, which shows the thickness of a water layer that would form if the rain fell on level ground.

Only four items are needed to make a gage: a large can, a tall, narrow bottle—an olive bottle if you have one—a centimeter ruler, and a permanent marker. After collecting your materials, work near a sink.

Pour water into the can until it is 1 cm deep.

Then pour the water into the bottle and mark the level on the outside. Write "1 cm" next to the mark.

Pour another cm of water from the can to the bottle and mark the level (2 cm) on the outside. Repeat this procedure for three and four centimeters.

Set your rain-gage can outside to catch the rain. When the storm is over, pour the collected water into the bottle. The marks on it will show the amount of rain, even though they are not 1 cm apart.

You will find that 2.5 cm is not an unusual amount to collect, unless you live in a dry state.

### SNOW GAGE

If it is snowing, let the snow fall into your rain-gage can. When the storm is over, just stick a centimeter ruler straight down inside, and you can tell how much snow has fallen.

Snow is really a frozen fluff. There is more air in it than water. Let the snow you have collected melt. Then pour the water into your rain-gage bottle and see how much there is. Or, rather, how little there is. Ten centimeters of snow usually melts down to only one centimeter of water.

Is the weather changing? Maybe you want to hurry outdoors without finishing your gage. There's a quicker and easier way to set up a weather station. You need only a one-meter rope. All you do is . . .

. . . hang the rope outside. When it gets wet, it is rainy. When the rope sways, it is windy.

# Jest Riddles

**Puns . . . Cornball . . . Zanies . . .**
And if that is what you like, these riddles are jest for you.
If not, read them anyway. Read them and groan.
And try to figure out:

What is over 60 years old, has over 60 legs, and weighs over 60 grams?

. . . A box of animal crackers—the kind that is 7 cm tall, 13 cm wide, 4½ cm deep.

The box first appeared in 1902, containing about 20 four-legged creatures, with a net weight of 2 ounces. Since then, the contents and box have remained the same. Now a change is due. The contents will be given in grams.

1. When is a 2-meter strip of wood not a strip?
2. When is an 80-centimeter tube not a tube?
3. When is a 13-centimeter bar of soap not a bar?
4. When is a girl weighing 55 kilograms not an ordinary girl?

1. When it's a board (aboard).

2. When it's a tire (attire).

3. When it's a wash (awash).

4. When she's a kin (akin).

Hippie the hippopotamus is 2 meters long and weighs 2 metric tons. But what is bigger than the hippopotamus and doesn't weigh a thing?

. . . The hippopotamus's shadow.

1. What happens in a 100-meter vampire race?

2. What did the ghosts want for their party that measured 24 × 30 centimeters?

**1. They finish tooth in neck.**

**2. Sheet music.**

1. What is the difference between a secretary using a typewriter and a half kilogram of sugar?

2. Some office workers press an elevator button with the thumb and others with the forefinger. Why do they do this?

1. One pounds away and the other weighs a pound.

2. To make the elevator go. Although the forefinger is about a centimeter longer than the thumb, either one works.

1. Everywhere in the world,
   An ounce is over 28 grams.
   And this holds true
   For rice and glue, beans and yams.

   Now, when 28 peas weigh a kilo,
   With all of them big and round,
   Can you tell us, please,
   The number of peas in a pound?

2. There's a synonym for
   A thousand-cubic-centimeter jar.
   But can you spell it
   Without using an r?

3. What is like the dinosaurs,
   That is destined to fail,
   That has no head or tail,
   That cannot change,
   And cannot bend,
   With a foot in the middle,
   And one at each end?

1. One **P**
2. **i-t** spells it.
3. A yardstick.

1. A pet-shop owner ordered a cupboard door 54 × 54 centimeters. When it was finished, the door was 56 × 56 centimeters. It was too big. When the carpenter cut some wood off, it was too little. But the carpenter cut some more off. Then it was just right. How did that happen?

2. How can you tell the owner of a shark from the other customers in a pet shop?

1. The first time, the carpenter cut off a strip 1 centimeter wide, which was too little. Then he cut off another strip 1 centimeter wide, and the door was just right.

2. The shark owner is the one buying a 500-meter leash.

When someone says, "There's a bear three meters long," what does it mean?

1. When is a kilogram of coffee like a hectare of land?

2. If it took 5 people one day to dig up 5 hectares of cabbage, how long would it take 7 people to dig up the same field?

1. When it is ground.

2. No time. It was done already.

What does
a 300-kilogram mouse say to
a 2-kilogram cat?

. . . HERE, KITTY, KITTY!

**1.** Here's a boy 160 centimeters tall, weighing 45 kilograms. What will make him fat?

**2.** Here's a girl 150 centimeters tall, weighing 38 kilograms. How can she get fat?

**3.** And here's a woman 165 centimeters tall, weighing 90 kilograms. How can she get fat?

1. If he is thrown up in the air, he'll come down... PLUMP

2. She can go to the butcher's.

3. She can't get fat. She's fat already.

In the jungles of Brazil, it is said, the mosquitoes are so large that many of them weigh a kilogram. They sit on logs and bark when people go by. How can this be?

... Wherever there are 10,000 mosquitoes, there are many mosquitoes. Now, if one mosquito weighs 100 milligrams, 10,000 mosquitoes weigh 1 kilogram. And, of course, people do see mosquitoes on logs, and sometimes on the bark of trees.

1. How can you lift a 2,000-kilogram elephant?
2. Why didn't the elephants care whether the lights in the zoo were 300 or 500 watts?
3. Why did the female elephant squirt 10 liters of water from her trunk?
4. Why don't elephants care about learning the metric system?

1. Put him on an acorn and wait twenty years.

2. They didn't live in the zoo; they were circus elephants.

3. There was a fire and she belonged to the elephant fire department.

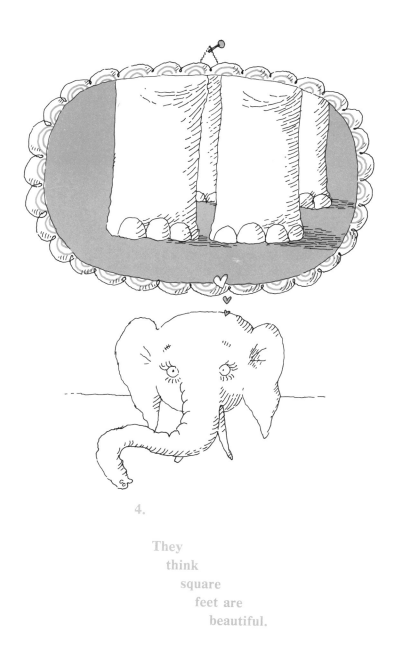

4.

They
 think
  square
   feet are
    beautiful.

Where can you find a kilometer that is only 1.2 centimeters long?

... Here.
In the
## INDEX.

angstrom, 20
calculator puzzlers, 90–93
calculator puzzlers, 90–93
Celsius, Anders, 27
Celsius scale, 27, 48, 64, 96; puzzlers, 63, 87
centimeter, as a unit, 6, 8, 19, 48, 99; cubic, 17, 25, 26, 53, 61, 67, 111; puzzlers, 7, 9, 10, 11, 12, 21, 45, 52, 93; riddles, 103, 107, 111, 113, 121, 127; ruler, 7, 8; square, 25; tape, 11, 12, 13, 20
conversion tables, Celsius scale to Fahrenheit, 83, 96; kilometers to miles, 55; metric to old units, 83; old units to metric, 82; pounds to kilograms, 14
customary system, 18, 29, 52
decameter, 19
decibel, 68
decimeter, 19
Fahrenheit degrees, 27
gigameter, 19
gram, as a unit; 22, 50, 71; puzzlers, 51, 53; riddles, 101, 111
hectare, 24
hectometer, 19
height and weight charts, 15
kilogram, as a unit, 14, 28, 71; puzzlers, 16, 49, 65, 76; riddles, 103, 107, 117, 119, 121, 123, 125; standard, 66
kilometer, as a unit, 6, 19, 20, 22, 33, 34, 55, 76, 78; puzzlers, 55, 77; riddle, 127

kilometers per hour, 54, 66, 97
kilowatt, 27
liter, as a unit, 17, 28, 62, 66; puzzlers, 17, 23, 26, 52, 61, 67, 93; riddle, 125
mass, 49, 53, 54
megameter, 19
meter, as a unit, 6, 19, 22, 34, 42, 76; puzzlers, 10, 56; riddles, 103, 105, 107, 115; ruler or tape, 10, 35, 38; square, 24, 25; standard, 28
metric system, changing to, 58, 61; prefixes, 22, 23; puzzlers, 25, 26, 47, 66; unit symbols, 22, 23, 27
micron, 20, 21, 22
milliliter, 53, 61, 84, 88; recipes using, 85, 86, 87, 88
millimeter, as a unit, 19, 20, 22, 48, 88, 89, 98; cubic, 25, 26; in making jewelry, 88, 89; square, 25
rain measurements, 98, 99
SI (Système International d'Unités), 30, 67, 68
snow measurement, 99
temperature, see Celsius scale
terameter, 19
time measurement, 27
watt, 27, 36, 125
Watt, James, 27, 36
weight, and height charts, 15; guessing, 16; pounds to kilograms, 14; puzzlers, 49, 53
wind speeds, 97

DATE DUE